Our Solar System

The Sky Above	2
The Solar System	4
Our Sun	6
The Night Sky	8
How Far Is It?	10
Glossary	12

Orlando   Austin   New York   San Diego   Toronto   London

Visit *The Learning Site!*
www.harcourtschool.com

The Sky Above

Look up! What can you see in the sky? The sun is in the sky every day. At night you might see stars. A **star** is a big ball of hot gases that gives off light and heat energy. Stars are in the sky during the day, too! The sky is just too bright for you to see them.

Each of the stars you see is a big ball of hot gas. Stars are far away from Earth.

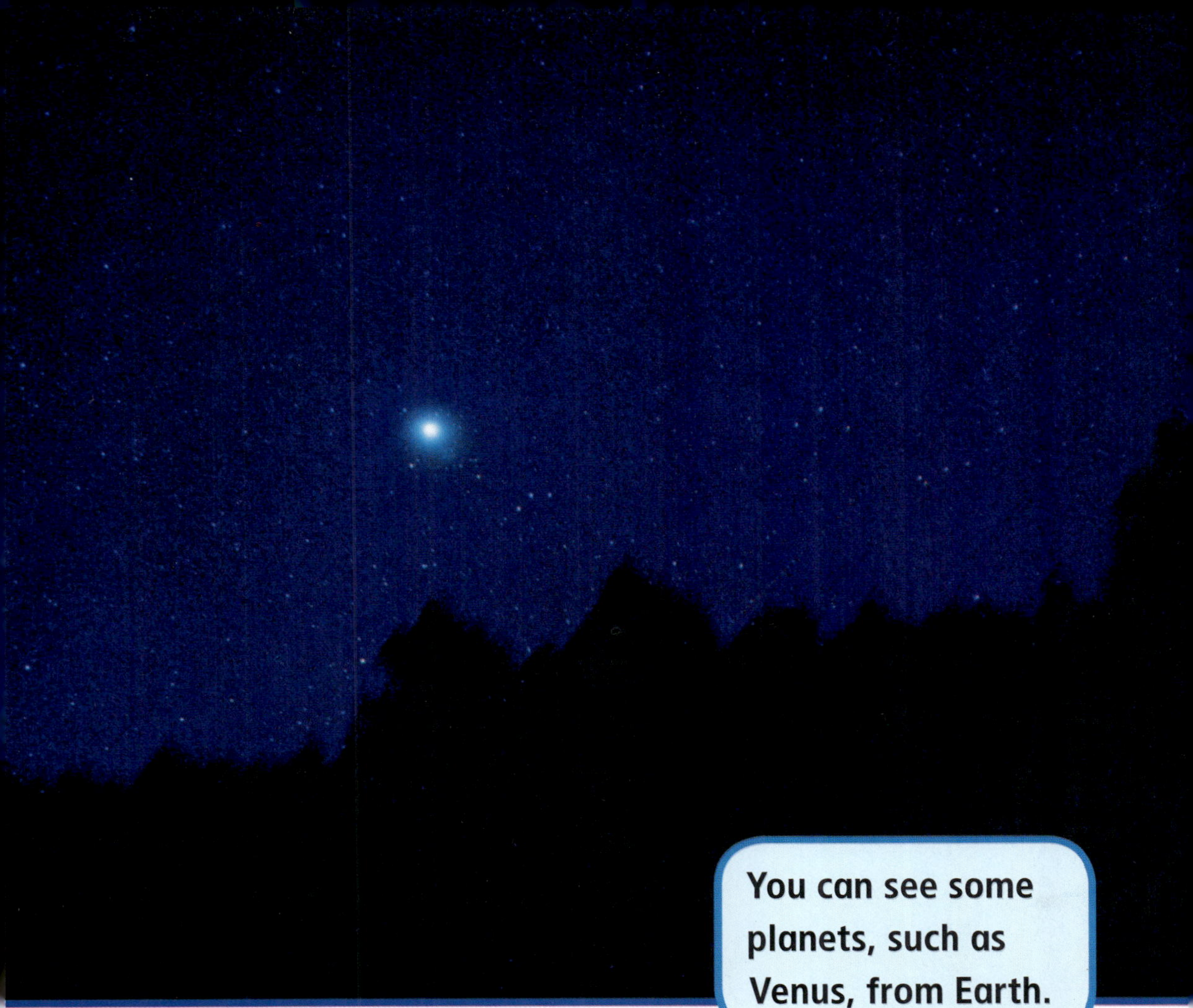

You can see some planets, such as Venus, from Earth.

You might also see a planet in the sky. A **planet** is a large ball of rock or gas that moves around the sun. Earth is a planet. From Earth, you can sometimes see other planets, such as Venus.

Focus Skill **MAIN IDEA AND DETAILS** What are three things you can see in the sky?

The Solar System

The **solar system** is made up of the sun, its planets, and the planets' moons. The **moon** is a huge ball of rock that moves around Earth. Our moon does not give off light. It reflects light from the sun.

Fast Fact

Many planets have more than one moon! The planet Neptune has at least 13 moons.

You see the sun and Earth. They are both part of our solar system.

Planets move in a path around the sun. An **orbit** is the path a planet takes as it moves around the sun. The path a moon takes around a planet is also an orbit. The moon is always orbiting. It appears to change shape as it moves.

Focus Skill **MAIN IDEA AND DETAILS** **What are two objects in the sky that have orbits?**

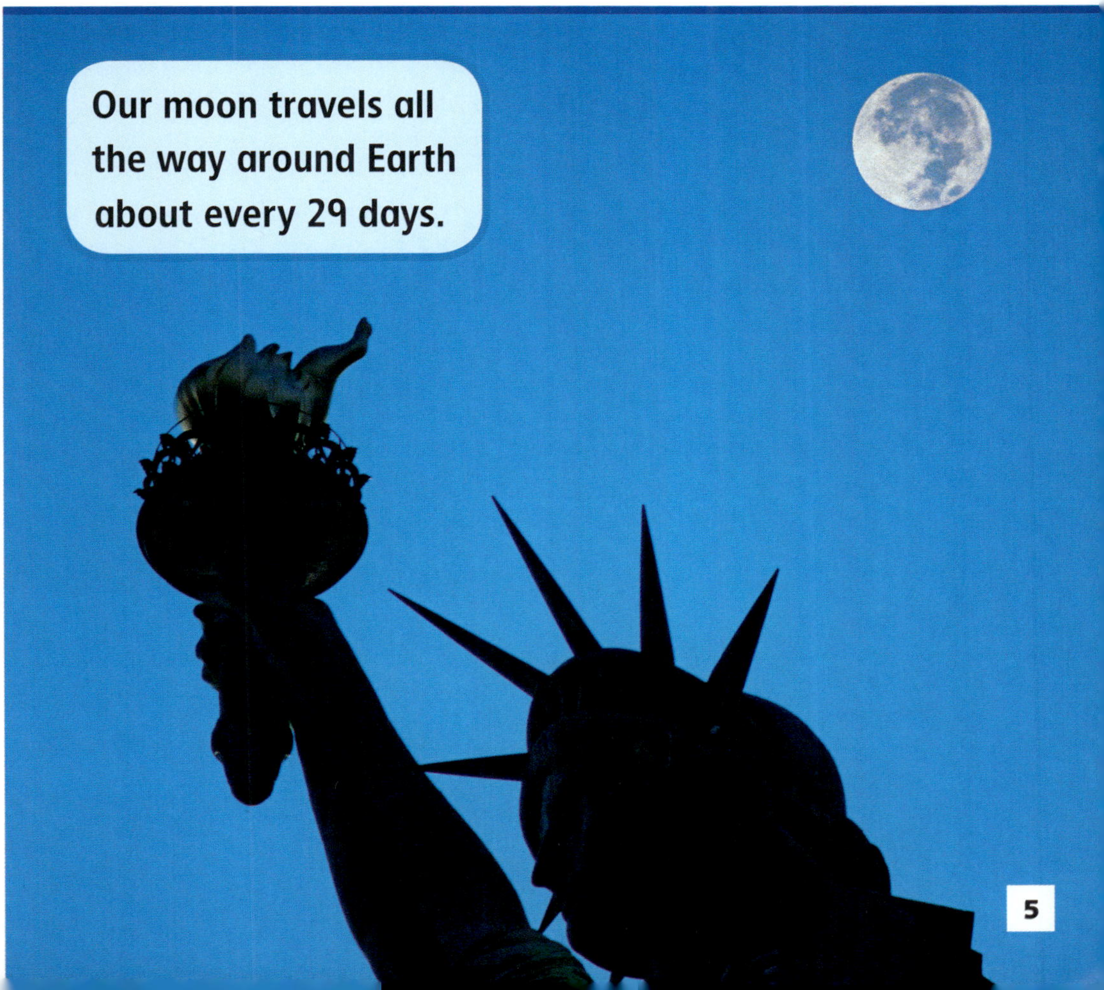

Our moon travels all the way around Earth about every 29 days.

Our Sun

Did you know the sun is a star? The sun is the closest star to the planet Earth. Earth orbits the sun. Earth gets light energy and heat energy from the sun.

You see the sun in the sky every clear day. The sun is the closest star to Earth.

During the day, the sun might be the only object you see in the sky. The stars, moon, and other planets are still in the sky. You cannot see them because the sun is so bright.

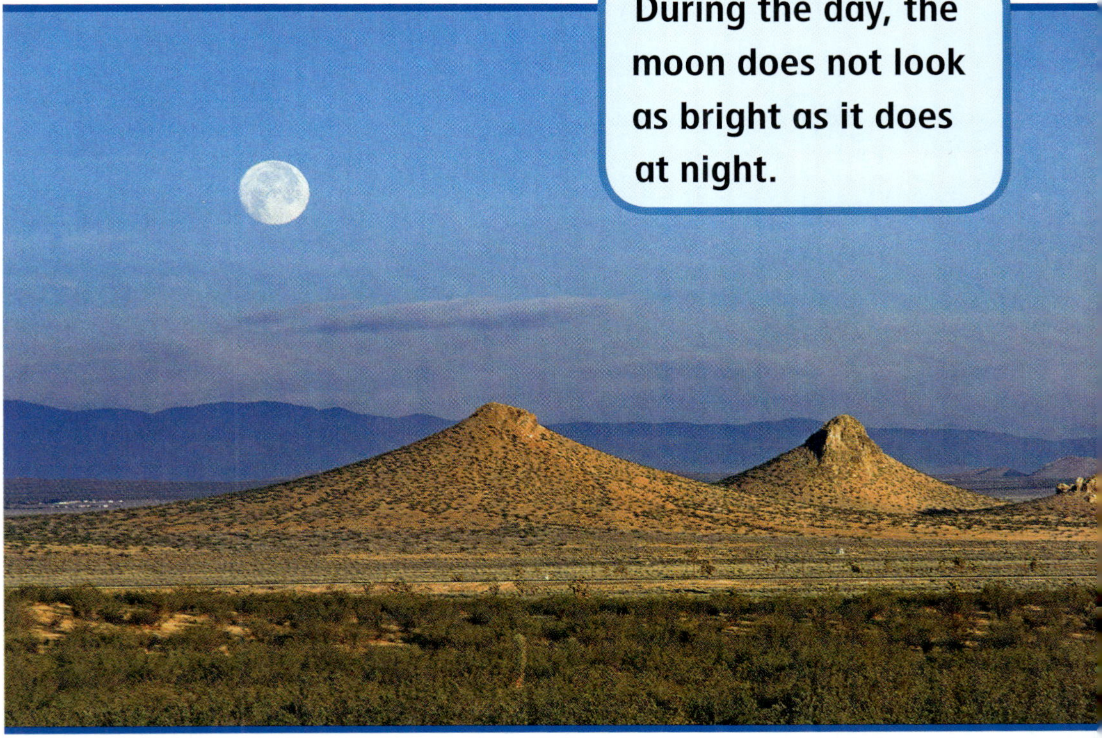

During the day, the moon does not look as bright as it does at night.

Sometimes you can see the moon during the day. It will not look as bright as it looks at night.

Focus Skill **CAUSE AND EFFECT** **Why don't you see stars in the sky during the day as you do at night?**

The Night Sky

You can see many stars in the night sky. These stars are like Earth's sun. Some of these stars are bigger than the sun. Others are smaller than the sun. The stars are much farther away from Earth than the sun. That's why they look so tiny. That's also why they do not look as bright.

Fast Fact

Sailors used the constellations to help them guide their ships.

The North Star looks very bright in the night sky. It is still very far away from Earth.

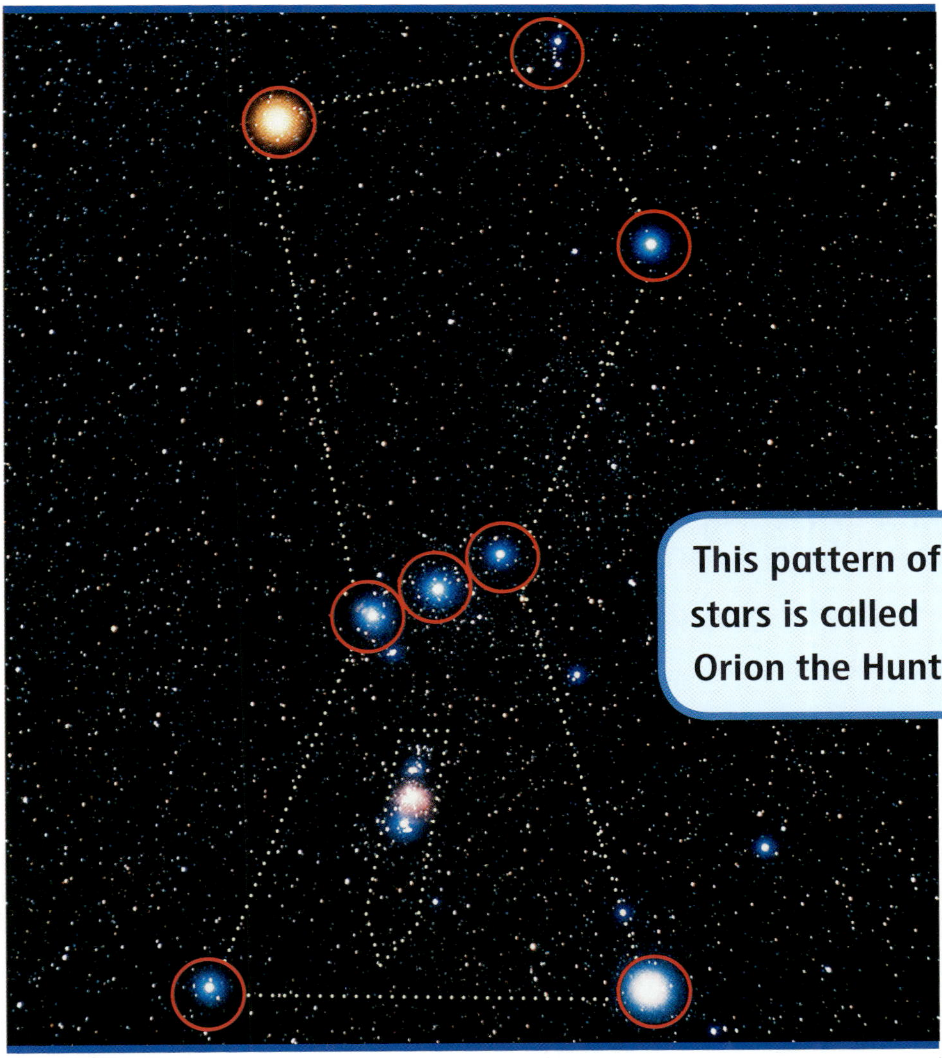
This pattern of stars is called Orion the Hunter.

A group of stars that forms a pattern is a **constellation**. Constellations are patterns made by connecting the stars with imaginary lines. Some constellations look like animals or people.

Focus Skill **MAIN IDEA AND DETAILS** **Why do the stars at night look very small?**

How Far Is It?

The sun is 150 million kilometers (93 million miles) from Earth. Two planets are closer to the sun than Earth. The other planets are much farther away from the sun.

The planets are different sizes. Venus and Earth are about the same size. Some planets are much larger than Earth. Other planets are much smaller.

Focus Skill **MAIN IDEA AND DETAILS** **What are some ways the planets are different from each other?**

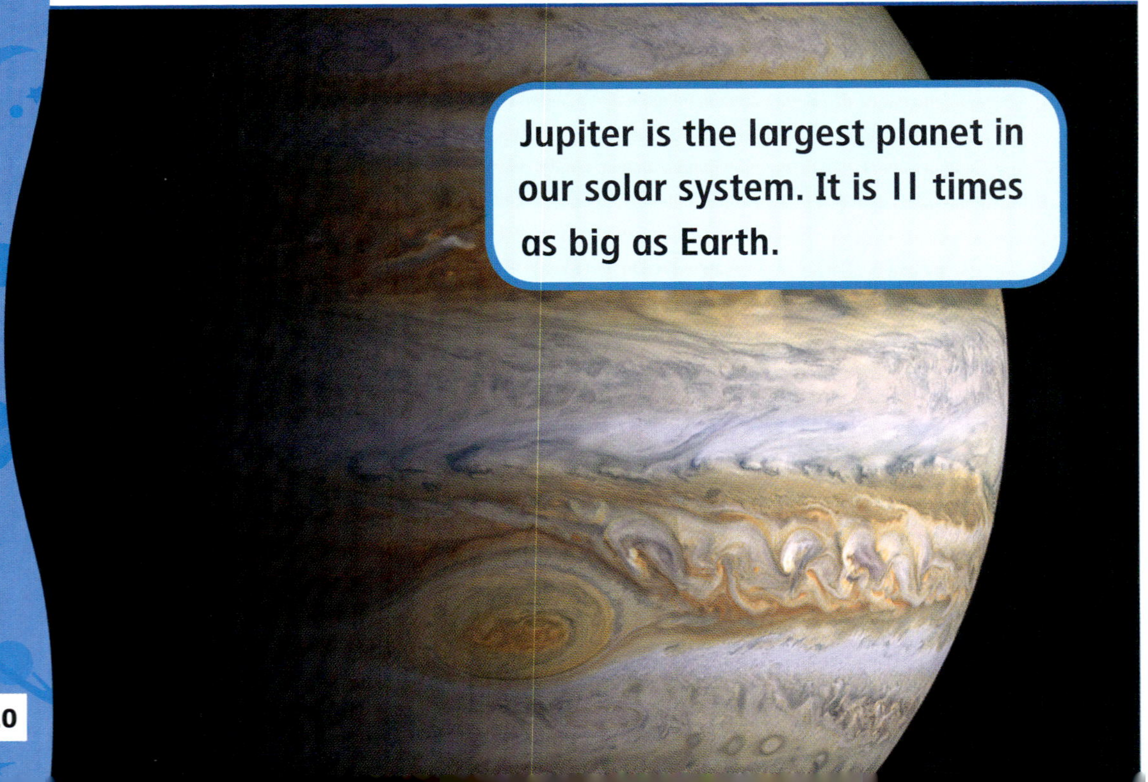

Jupiter is the largest planet in our solar system. It is 11 times as big as Earth.

This is how Earth looks from space.

Summary

The solar system is made up of the sun, planets, and planets' moons. Planets orbit the sun. Moons orbit many of the planets. The sun is the closest star to Earth. The sun's brightness makes it hard to see other objects in the sky during the day.

Glossary

constellation A group of stars that forms a pattern (9)

moon A huge ball of rock that orbits the Earth. The moon takes about 29 days to go all the way around Earth. (4, 5, 7, 11)

orbit The path a planet takes as it moves around the sun. Earth's orbit around the sun takes one year. (5, 6, 11)

planet A large ball of rock or gas that moves around the sun. (3, 4, 5, 6, 7, 10, 11)

solar system The sun, its planets, and the planets' moons. In the solar system, the planets move around the sun. (4, 10, 11)

star A big ball of hot gases that give off light and heat energy. The sun is the closest star to Earth. (2, 6, 7, 8, 9, 11)